1

Also by Jack Tyler Jones

Vice and Virtue

The Island of Killing Tourists

Adam at the End of the World

Author's Note

This was written in a single afternoon, as an experiment, while engaging in substance use.
Therefore, typographical errors have been left in.

It remains how it came.

They call me Dream. It was October 9th, when I stepped onto the bus, out of life as I know it, and into a spiral toward sex and death. I wouldn't have expected it, necessarily, but that's how these things work, sometimes. It's not your watch not going off. It's the extra four minutes you took in the shower, and two for the coffee, and three for news, and two to pack lunch, and one to wonder whether you've forgotten something. And you have: you have forgotten life as it once was, by making yourself late.

She was there on the bus, naked. Not actually naked. There was a red dress, but the red dress was inconsequential. She had a book on her lap, and the book was face down. One leg was draped over the other, like fallen pillars concealing an ancient cave. I knew I'd be there, even before I realized that I wanted to be. Destiny just rushed up and punched me in the throat, which is an excellent way to make me try and avoid it. I don't know how I know things. Sometimes they just arrive in my mind.

I trailed my gaze up her body in exactly the way that every woman hates, and hated myself for it. There were two small brains resting under the fabric where her breasts should have been; I cannot picture a woman's breasts without first getting to know her brains. It is unfortunate but unavoidable.

There was a handshake coming toward me on one end of her arm. Her arm which, heretofore, had been draped, like the other one, across the seatback beside her. And there it was: Fate, creeping around my legs like a python, stealing me for the things that were about to happen. This, I knew, ten minutes late, taking the bus instead of the train I missed, this was it. This was that unfortunate, unstoppable thing that was going to make me die. I had set it in motion. I just had to linger over the scent of the conditioner, didn't I? Had to run it through my hair, which I ought to have just cut off and been done with. But that's how these things happen. You don't know, when you buy the conditioner, that that's what's going to lead to your demise, by making you late. No sir. No ma'am.

It's sensational, really, that handshake. "My name is Fidelity," she says, and right then, I think that I am wrong. I think that, no, this is not the unstoppable force of destiny coming to claim me in the form of a mostly naked woman. This is a prostitute who does not yet understand New York. But I am wrong about being wrong. I see her forty eight times on the bus between October and March, enough times to realize that I am making myself late on purpose, that oh my goodness, I want to sabotage my solitude so subtly that I'm blindfolding myself against my intentions.

"That book," I say, on the train. Because she has begun to appear on trains. "You always have it."

"I do," she says. And with that statement, and the smile that comes through her red lipstick – either a threat or an invitation; it is impossible to tell – I know that she has married herself to me. I cannot escape. I do not wish to.

Fidelity makes her way past my office windows, floating by on her feet when I know perfectly well that she isn't there. She waves her fingers at me.

"What do you have written in that book?" I ask her, on our third date, at the café. I have asked her to be my companion for the evening on two previous evenings. It is not me who wanted this, really. It was the force acting upon us both as puppets. She sits at the end of her cosmic strings, not knowing this. To her, I am just a man passing on a bus so frequently that I eventually asked her out. Maybe we were both bored, and out of other options, she tells herself, behind her eyes. Maybe we were not thrust in to this dirty blue planet just to come out holding hands.

She is wrong. I smile at her, her fingers trailing down the leathery spine of her book. "Oh, things," she says, shy, and daring, giving me the kind of look that gives one the feeling of tumbling down a great large hill.

We have a slow dinner. It lasts for five spiraling autumns, though only forty-five minutes have passed. I can feel myself growing older in her eyes, older and odder, until she accepts my oddness as familiarity.

"Press on," she says, with a tinkling laugh, when I insist on paying for our meal. But it is there, burrowing a hole into my wallet: love, that great facilitator of bad decision making.

"How far would you like me to press?" I say.

She considers me, like an elevator considers a building, up and down, from the inside. Her look was so penetrating, I mistook it for my own reflection in a well. She shimmered. "I would like you to press as far as your apartment. Maybe wine. Maybe further. Maybe I will do some pressing."

I do not press further. I attempt to backpedal, to retrace. I am not out to get you out of your clothes, I want to tell her. Oh, yes you are, her eyes tell me.

I will kiss you goodnight, and then walk home, my actions say.

She does not speak to me on the train for four long months. I miss and catch buses, trying to catch her, trying to make my mistakes unhappen. Rejection curls its long, wayward fingers over my heart, plowing that ground until it's fully open, no wound left to be inflicted.

It is then that she is poised for her return.

"Poems," she says, on the bus across from me, totally unprovoked. "That is the answer to your question. I write poems in the book."

"Poems. And do you intend to publish them?"

"Publish them? I am them."

We walked down the cobblestones for four evenings, not saying anything, but not not saying anything, either. Then in two months, I noticed the way she changed the coffee filter differently than I did. It wasn't natural.

"We are natural," she said. "We are the environment. I am your habitat. So aside with the coffee, and its divisive pathways. Inhabit me."

I inhabited her on the floor, next to the pantry. She kept her shoes in there. She had a pair of turquoise ones. This always bothered me. I didn't know why, until one day, I did.

"You never wear those shoes."

"You never asked me to."

I could see, then, that she would not make this easy, this beginning to coexist with another, this asking her if she never wore the shoes, and yet the shoes always appeared to be worn, and wearing down, because, perhaps, she is someone else in the daytime. Someone else being somewhere else doing something else, with people and places and things that aren't me.

It was this unsettling thought that settled me down. "Audrey," I said, because sometimes I called her Audrey, because I didn't know her name, "Audrey, will you wear the turquoise shoes?"

She was a librarian, a keeper of worlds. I respected that. I didn't know it about her, until the day that I asked her to put on the shoes, and followed her to her job, which was not like my job, which is red, and high, in an unpleasant building. The chairs have the capacity to spin, but the people inside of the chairs do not. And only those who would spin, perhaps, ought to be in chairs that can.

There were children at the library, and the children called her 'Miss.' I found this quaint. I too wanted to be like one of the children, so I sat on the rug, to listen to her read this story, but she whispered to me, off to the side, that this would not be permitted, that the parents would not understand. I left the library.

I bought Audrey a cupcake. I was still going to refer to her as Audrey. It's what we do to people. We discover that they are librarians with shoes unlike the red ones they had on when we met them. They are inhabiting stories other than our own. We find that hard to grasp.

She grasped me when she got home that night. "You left without me in your pocket," she says, retrieving my cell phone from her purse. She plopped it into me, that way of communicating, and I looked back at her, and I said, "Audrey."

And she said, "Why do you call me that?"

And I said, "Because I don't believe Fidelity. No one names a child that."

"But I am not a child."

For three months she came floating by my windows, with her hair blowing, when I knew she wasn't there. And now I even knew where she would be, knew that she would be opening little paper doors for children to walk through into other worlds.

She opened one of the little paper doors for me. "I wrote this. The day after we first met, on the train." It was a poem, and it wasn't any good. I remembered each word the way one remembers turns on the road going home. Each line ended in a clanging, upward rhyme. She was doing something to me, some kind of unfit sorcery, because as I read, and as I saw how bad it was, and how simultaneously she must have pined over it, stroked its every poor metaphor and nuance until they came away slippery in her hands, that destiny had well and truly taken a hold of me. The truth was not available to be told, and a man like me never lies.

"You are a lioness of the page," I managed. Only, one does not manage. One only 'says.' And to say and to manage are drastically not the same.

"You have a forked tongue." She pointed this out to me two days later. "Why don't you hide it from me?"

I was not in the habit of hiding things. "Men in fedoras," I said, "are not in the habit of hiding things." I had insulted her mother by some method unknown to me, but was fully prepared that day, for no other reason than it being Monday, to be rid of her, and all of her bullcrap, too.

"You are a magnificent destroyer of false hope," she said.

It was my job to be a magnificent destroyer of false hope. I had to shake down every false notion of me that she had, and of what the false mes might potentially give her.

"I am not as important to me as you are important to me," I told her. I felt that this was dramatic, but that's where dramatic things are to be said, in the kitchen, aren't they? This carried on. There were things thrown. A saucepan did not break, but only because it was a very strong saucepan. My head broke open, and at first I thought that she had hit me with the saucepan. But it turned out that my head had just broken open on its own. These things happen, sometimes.

My brain spilled over the front of her dress, which was suddenly the same red dress as the first time, that day on the bus, when she wasn't wearing anything. I considered the breasts where once I had been able to imagine only two small brains. "I have come to know you," I said, in the moment that I realized she had always been naked.

"There are no turquoise shoes, are there, darling?"

She backed away from me slowly. "Yes, dear. There are. What are you on about?"

I was on about nothing. It was just that I had decided to start a new career, as a dentist. I began to wear the white coat everywhere. It's what dentists do. I embraced teeth. I felt like the teeth were embracing me, as well, in the books I read. Each word typed was also spoken by a mouth, was it not? So I felt very connected to the world, and I no longer wished to remove its teeth, though that also meant that I couldn't clean them anymore.

Fidelity took me away to the country, to breathe in the fresh air, which she said would be good for me. My need for her had been unmasked, terribly, in the way that needs have a way of unmasking themselves, when I told her that I wasn't crazy.

"No, my love, of course you are. Crazy is what makes you an artist."

That was when I knew that the dentist's coat was a painter's smock, but this frightened me. I did not want to be a dentist or a painter. "Let's go back to it being like in the mountains, but back where the buildings are, back where you love me, before this." I tell her this in a lightning bolt of realization. She escorts me back to the café where I find that, bizarrely, we are still on our third date, only the third date.

"You have a magnificent way of speaking," she tells me. The girl could lie the spots off a dog. Or could she? I couldn't tell, and that's the scary thing about liars: you never know which time period you're in, or even if, at that point, they really are your wife.

I shake my head. I can't have a wife. I've only just seen her on the train five minutes ago.

"Miss," I say. "I believe you've dropped your book."

"Thank you."

She gets up and crosses to the other end of the train car. It's just now when I realize that there is no book in her hands, it's just the one here on my lap, the one I've been writing. I thought we were on a bus. But we are on a train. And I do not appear to be late for work.

I glance at my watch. Where have the last fifteen minutes gone?

The woman walks back down the subway car toward me. Great whiskers, it's a subway now? Where has my reason gone? How did I come to be under the ground?

The red dress falls off of her as she approaches me. There are no other people on the train, only cats. I'm unaware of where the cats appeared from.

"I'm not sure if I know who you are at the moment," I confess.

She is arraigned briefly in the clothes of a priest, and I see her with a white beard, and the face of an old man. The church rushes past us, line by line. Passengers get on and off of the church. "I told you," she says. "My name is Fidelity."

I scratch my head. This is all most puzzling. Or is it almost puzzling? Because if there is one thing that a true puzzle ever does, then that thing is to bring you as close as you can get to other puzzles, and then leaving you to wonder what happened.

I was standing in the bottom of a canyon. There were things rushing past me, and some of the things were seasons, but others of the things were outer space, and I didn't know how to interpret them.

"Take your words and swallow them down," Audrey said. I was lying on the kitchen floor. It was right now, and there was a purple pile of dust in her hand.

"I don't think this is how this is supposed to work," I said, as the wall cascaded into clocks. Audrey wasn't Audrey any more.

We were having dinner. I wasn't sure whether or not we'd met yet, but maybe it was about to happen. We were outside, on a bus.

"This is perfectly normal," she assured me. Her legs were stretched open. I wasn't sure what to do with the space between them; I certainly wasn't big enough to fill it. "It's just an appetite we women have, from time to time." She told me this while stroking a panda bear. I didn't understand where it had come from, but I supposed it had come from China, and so I supposed I didn't care.

"Isn't communism important to you?" She questioned me often. Usually in another language. "I think we ought to take down some notes and waterfalls."

"Waterfalls," I agreed, because I had no idea what she was saying.

We were stranded in a jungle at the back of my head. I hadn't ever explored this part of me before, and I wasn't sure if she was allowed to be there. "The cats on the bus," she said. "They were really just the shadows of the cats here, big jungle cats. You ought to let them eat you."

And I did. One came up to us, and it swallowed me whole. I waved goodbye to her as I fell down its throat. I was relatively unsurprised, even bored, to find her there at the bottom, in its stomach, waiting for me.

"I'm going to try to leave you," she said, and then didn't.

The lights were spinning. The cat's stomach was a stage, and I didn't know why I was playing the guitar, because I had not touched one in a few years, and when I had touched one, it was because she was trying to teach me how to use it. Audrey was a teacher. I had forgotten that. She was my favorite teacher, in the fourth grade.

"I'm your favorite teacher now," she corrected me. I supposed that, since she was a teacher, I stood corrected, and that was all that mattered.

"Do you suppose we can just carry on now?" she asked. She was holding a clipboard. I didn't know why. "We need to finish checking up on the monkeys."

But there wasn't even a jungle anymore, so I didn't think that that made sense, until I noticed the lab coats, and the fact that we were standing in a lab. "My god, can't you even remember the groceries?"

I looked down the long line of monkey cages. "I'm not your god."

"And it's a good thing, too! I'd nearly forgotten. So sorry, my love." So we went to a temple. It seemed like the next logical step. But when we arrived the steps had just lead back into the same kitchen where she'd been standing over me, telling me to take the glass of water, which was a magnificent lake.

"Down to the Late Lake, or we're going to be late!" she called, running down a teal path. I suspected that the path had something to do with the shoes, the shoes she'd had when she was being a librarian, which might or might not have been a trick.

"I am really tired of this," she sighed.

When she sighed, I came out of her. I hadn't realized that I was a woman's breath, held in. It's amazing, the things that we don't notice about ourselves.

"Your mother is never going to see this, yes she is."

"Why did you say that sentence like that, Audrey?"

"My darling, are you sure that I've said anything at all?"

I decided to have a cup of tea with the number 13, to try and sort all of this out. I kept seeing Audrey's reflection in the stuff dripping down from the tiger fangs. Who was she, anyway?

"My name, for the third time, is Fidelity."

I suppose I have a problem with that, but the problem isn't worth mentioning, because Fidelity will never love me in the first place.

"Oh yes I will," I hear her thinking.

"Your thoughts were louder when you were a cheerleader," I commented. I wasn't sure if I had typed this, or spoken it.

"Louder but fewer," she typed into my head.

It was only just then that I was the piece of paper in the woman's book. She had been using a typewriter, in a café, to write stories about me, only I was the story she wasn't sure whether or not she wanted to keep telling herself. So she folded me up and stuck me in the little side pocket of her overcoat, which was a small leather book on my lap.

"Are you certain?" she asked me, and I asked her, or we asked each other. I wasn't sure which.

"Nothing is what we tell ourselves that we're doing. But you're telling yourself something that might not be the truth."

"The truth is the only thing," she insisted, pasta sauce dripping down over her nakedness, making a red dress. "There aren't any lies, except for when people make them."

"We should unmake the people," my fist said to the table. It hurt.

"We can't do that," I reasoned, or she reasoned with me.

"Stop being so different from me," she said, before abruptly walking into my chest and disappearing for two months.

I was having coffee by myself. I didn't like coffee, but I knew that she did, and I thought that this might draw her out of me.

"I am a space," she said. "You keep quoting me as if I am a person."

"If you are a space, then where are you?"

"I am not a question of where I am. I am a question of what you are."

"I am not a what. I am a who."

"Congratulations. Moving on now. Rocks."

"Why are we talking about rocks?"

"Because we are rocks," she pointed out, and I was mildly disturbed to see that we were. We were rocks, changing into a rock concert. It was then that I became a sweaty shirt sticking to a teenagers back, and I did not like this, so I said, "Teenager, there are better ways to get sweaty."

And she turned around, and it was Audrey, and she said, "I am older than you."

And I said, "But I made you."

And she said, "No, you only loved me."

And I said, "That's the same thing."

She burst into stars. They fell all over my kitchen, and my skin, and the bus. They were globular, and hot, and wet, but the wetness was not hot. It was frigid, ice cold.

"Why didn't you bring penguins?" she asked, as we stood up in the arctic, brushing snow off of ourselves.

"I didn't know we were going to be scientists next. Do you have the binoculars?"

"Nevermind. It's getting warm. Go bird watching." She appeared to be sucked backwards, although I had not yet put the binoculars up to my eyes. She was a bird, though, I noticed that.

"Why are you a bird, Audrey?"

"Why don't you stop calling me Audrey?"

"Because I don't like the suggestion behind your name. It tells my mind too many things that my mind does not go on to see as unpretentious."

She took hold of my penis, then, which I wasn't sure that I had, and pulled it out from under my red coat. That's funny. I don't remember siding with King George's men before.

"You weren't British before," she points out.

Have I become British? I suppose not, since I have just coughed up Australia. It is wet and folded over by my feet.

"You have things to be concerned about," she told me.

"So do you," I challenged her. She shrugged, melted, and oozed through the floor.

"Don't leave me," I worried.

She dripped back down through the ceiling. "I haven't. It's you who ought to leave. Get out of my house."

She pushed me out of a front door, and down into a southward facing attic. "There are important papers in the trunks," she assured me. "Just ignore them."

"I suppose I ought to pack this suitecase?"

"That would imply that we are going on a trip, you pretentious fist of fishing line. Untraintrack yourself."

It was then that I knew for sure we'd never met. I'd just had a headache, that was all. It just meant that I needed to take my yesterday pills.

"There aren't any pills," she insisted. "Pills are for people who are doing it wrong, and you aren't doing me at all."

So I fixed that, because when something needs fixing, sex is usually the superglue.

"You are just one big pile of mistakes," she whispered, reiterating everything I most suspected that other people were secretly thinking about themselves.

"This is not interesting," she insisted, opening up a whales mouth, and pushing me into it and pulling me out of it at the same time. "I think that I will just be you for a little while, because you don't seem to be doing a very good job of it yourself."

"I don't think you ought to be too concerned about that," she said, through my mouth, which was trying to contradict her.

"Go ahead then," she challenged. "Unthink me."

I was neither able nor willing nor willing to be able to do so.

"What an irresponsible man you are," she said, when she was me, looking down at her own body. "It is strange, having a flat chest and a penis, but not nearly so strange as it is for you, I bet, having the mind of a woman."

And then she left me, abruptly. I hadn't ordered meatballs. The food hadn't come yet. The spill, the spill of the red sauce, dripping its way down her cleavage, and into my imagination, hadn't happened yet. I was getting ahead of time again.

"Have you fallen down this well before?" she asked, taking my hand. We were standing on the edges of our forks.

"I don't see a well," I hesitated to say.

"I don't see an unwell," she said, and then jumped.

She fell, and it was a long time before the gravity on her fingers caught up to mine, pulling me down toward her, and through her. I had never been anybody else, before.

But there I was, sitting and looking out through the brown around the pupils of a man who I knew, from the inside of his head, to be vaguely familiar. He was always worried about his work, this man, and whether or not other people thought he was working too much, or not enough, to earn income for people who he thought would pop into existence once he reached certain landmarks in his life. They were probably arbitrary landmarks, things like school, and a job, but he had a job, and his job would crawl into bed with him at night, and sometimes it would comfort him with the cool, slender body of his laptop, and other times, it left him lonely, and it left him broken.

"You ought to record me," I said to the man whose head I was inside of. He was not at all interested in this suggestion. He was, indeed, not interested in the slightest bit of anything I had to say. I carried on for a day and a half inside of his head like that, until he took himself to a doctor.

"I think I've gone mad," he said.

"Fine. I'll leave."

The doctor was Audrey. She had said that she was going to leave. "You really mustn't do that," I implored her.

"Then you really mustn't wander off into other people's heads." Her verbs smacked between my jaws, delicious and crispy.

"Cook me up another batch of your words," I requested.

"Would you like them adjectives, nouns, colored and deep fried?"

"I want only the most fattening and unhealthy and marvelous of your suggestions."

"I suggest that you get up off of the floor," she said. She was made of tiles, black and white, and they were underneath me. "You've got a long way to go."

"Fidelity," I called, without saying anything. "Come back here."

She had left me standing on a long, winding road, straight out in front of me, with nothing around it but thousands of corners and shadows and questions. "Don't you love me anymore?"

"The sun does not love the earth," she said, sounding both exasperated and patient. "The earth merely goes around the sun. Figure it out."

I didn't like being someone else's religious pouting. She was starting to look like that old man again. "Audrey, trim your beard," I asked her, because she was now 300 years old, and had been debated by theologians.

"Don't tell me what to do," she said, without words, because she was laughing, and licking my face. I was alarmed to find that she was a cat, one of the cats from the subway. She was too small, and to uncatlike, to be a cat, though, and she wasn't licking, no, not anymore, she was nibbling my shoes, a rat, a fat sewer rat.

"I'm not fat, and I don't stink, I do not," she protested.

"Of course you don't. We were in the bathroom. This was when she was me, when I was being a young girl. I did not like that at all, even less than the man with the brown around his pupils liked having someone else occupying his headspace with him.

"Don't apologize to me," he said. "I came back here only to tell you that I no longer resent you for abolishing my ties to the natural world." He then tipped his hat at me, and with his briefcase, he spun around once, and became a woman in a red dress.

"Let's go to meet me, on the bus," she suggested. "I will wear nothing, and you will be a man unable to know what the nothing looks like, and hating it, as men so often do."

"I don't know why you want to play your cards this way," I told her.

"That's just the way card games get played. You ought to have chosen chess."

"Chess is for beginners," a knight said. We were on his board, in his thought experiment, and it was a long time ago, maybe the 1300s. I recalled that I was supposed to be having a cup of tea with the number 13.

"Finally," Fidelity said, throwing up her hands. "You've remembered that nothing gets forgotten."

I was just beginning to wonder whether aliens existed, but I could feel her impatience as she tapped her red fingernails against the book on my lap, which she'd been writing in. "You are a stranger, and you are stranger all the time."

"You've done it again." This thought came through me, an echo. "You've mistaken me for my sister, Consistency. She's a twin who died long ago. Let her be the thing that isn't."

I sat down and cried, because the funeral going on around me had just become apparent. There were shadows and cobwebs of disappointment hanging everywhere on the tatters of things.

"Everything that can't wait for a few hours is going on without you anyway," she said. She wasn't angry.

"Come into the cave."

There wasn't any space there, because there weren't any walls. There weren't any edges. "Keep turning on the lights," she coaxed. "Come down the hallway."

I saw our lab coats again. We were in the place with the chimpanzees. I was uncomfortably aware of being the chimpanzees themselves, as I saw her beauty reflected in my large, brown eyes.

The man in front of us, in the lab coat, ran his hand along his face, lengthways. "I have started to come undone," he said to himself.

His wife was standing beside him. Her name was Audrey. I hated the fact that her name was Audrey.

"Why did I ever start calling you that?" I lamented.

"Don't 'lament,'" she said. "It upsets the things that aren't."

"Everything worth weeping for has already happened," she said, from inside of my voice, with the accent and inflection of a man.

"You've developed a habit of saying yes," I said to her. "This is our third date, this thing at the café. Which one of us has the book?"

"I wrote you a poem," she said again. "From that day on a train."

"But we met on a bus." She didn't like where this was going. "We haven't met yet."

I could feel her getting younger, impetuous. "I forgot that. Sorry dear."

"Shall we go and have our first kiss?"

I wasn't sure which of us had made the suggestion, but it was a good one. We had our first kiss in the moment that one mouth slit off to become to people. It probably happened a long time ago. Everything will look like a long time ago, when you're young, like me, looking up at your old age.

Old age zipped away into a tiny black spot. She said something, but I didn't know what it was. And then she disappeared.

"Fidelity?" I said, into the light.

"Yes."

There were a great many yeses going on in those days. It was when we had the castle, before we'd known what it was like to not be middle aged in the Middle Ages. She was exhausted of working the plow, so she turned herself into several servants.

The servants began to argue with one another about which of them ought to die first. Hunger won, but he died, because Thirst and Breathing ought to have won.

"Sleep is sometimes the most important thing, but Audrey," she said to herself, out loud, "You need to stop being Audrey."

I thought about my name, and why they called me that. I had assumed I was a lion, but if so, then why was I bunch of mice, in cages?"

"Get me out of this lab," I said, as a chimp, to the chimp in the cage with me.

"Look up at the doctor." It was Audrey's voice, coming at me from inside of the chimp. "I can't call you Audrey anymore," I said, telepathically, to the woman in the lab coat standing next to the puzzled man.

"It was never my name in the first place," she told me, gently. "And you haven't gone anywhere." She had anticipated this thought, just as I was having it. Only, I had gone somewhere, or rather, I was going there, and I couldn't decided whether or not I liked it.

"That's because decisions are destinations," she pointed out. "And you like traveling. Just leave the lab on your own time."

I walked out by blinking. I was a colony of monarch butterflies, and we were in South America, but I suspected that we could go and see the real monarch, in Britain, very quickly if we wanted to.

"She'll be with you yesterday," Audrey commented, patting me on the head, which wasn't there. It was just an open space where a head used to be, before mine burst open, spilling on to her shirt, that morning in the apartment, when we had disagreed about the coffee filters.

"You can change the filters however you want," I said passionately, and she made fun of me for saying it "passionately," and I almost hated her for that, but then I didn't. I had decided to like her, because liking people felt better, even if the people didn't like me.

"The people ARE you," Fidelity said. She didn't usually speak in capital letters. Her lipstick was on, but I could see it making its way back into the tube as the day edged toward sunset.

"Right. We met yesterday."

"Two weeks ago, on the train, but also on yesterday's bus. This is our third date."

"If you say so."

"Who says so?"

"Pardon me ma'am, have we met?"

A gentleman appeared with a menu. "Is this gentleman bothering you?" he asked us, apparently about himself.

"Not in the slightest," I informed him. The man bustled off. I suspected that he may have taken our plates with him as he went.

I found myself standing on a plate, surrounded by a dinner that I was going to eat in three years time. I knew that I was going to be eating it with Audrey, or Fidelity, or whatever her name was, and I knew that, three years from now, having that dinner, I would know her name.

"In the future, I'm a man," she reminded me.

"No, I'm a man. I finished up with being a little boy last Tuesday."

"I think I should like to pay a visit to last Tuesday, then."

I recalled babysitting her. I recalled her babysitting me. It's an unnatural business, being parent to other people's children. "But we are our children," she corrected me. "There are no other people."

Thoughts, in that moment, seemed inherently incestuous, as they were always breeding in families: worry, fear, doubt.

"What you need," she cautioned, drifting down through another ceiling, "is to be a little less cautious."

"Your day off!" she admonished me. I had been thinking, wondering where it was that I was trying to get to, in such a leisurely hurry, that had allowed me to be a full ten minutes late, causing me to miss my train, and see her there, for the first time, on the bus.

"You see virginity like a second chance. But there can't ever be. There's just a hole. It was no less of a hole before or after filling."

"But a hole, once filled, can be unfilled."

"But it can never unknow the spaces that it has been for other people. The opportunities that it has created, and missed. People fell into that hole, waiting for it to become some type of theological argument, but it was only ever virginity."

"There are people who think that God came from virginity."

"Did he?"

She sat with me across the table in the café, but I knew then that the café was just a ruse, and so was she. "Fidelity," I said. "Speak to me honestly."

"I've never been no other way," she said, with a sudden Southern drawl, completely unlike anything that she had ever said before.

41

"The important question about God," we told each other, "is not 'Did he?' but it's always, 'Is he?'"

Is he like this, or is he like that?

"You're finding out all the time, you champion," she congratulated me.

"This is absurd."

She shrugged. "Only if you think so."

She was a cuckoo's nest for a half an hour after that, and for a half an our, as well, because there was only half of me without her, half an hour as half of the our.

I very much wanted the company of a gentleman, such as myself, so I went to Oxford. I put on my hat and coat and pipe, so that others would mistake me for Sherlock Holmes. People were always mistaking me for Sherlock Holmes; whyever else would they ask so many questions?

When I looked down to see if I'd spilled anything on myself, which seems to happen quite a lot these days, I noticed that where my body had been, there were now only question marks.

"How inconvenient," I remarked, to no one in particular.

No One In Particular had just finished sowing herself into a red dress. "Well, if that's all you care for it," she said, throwing up her hands. The dress melted off into a puddle of red sauce, which might have been blood. I decided to ask myself, since I was a detective.

"The source of the malady is more chimps in cages," the woman in the lab coat said to me. "But that isn't true. It's just my job to stand here and say so, until you give me a new job."

"I should very much like to present you with a new job right now," I said, mopping up the blood from where I'd killed her. She formed again when I squeezed the blood out of the mop.

"Always the same DNA, always the same idea," she assured me. "I'm everyone else but you."

I then understood, perfectly, that the world was a matter of lunatics. The fact that I would no longer think so by the time I finished typing this sentence was proof of the matter.

A large stack of mattresses fell out of the sky. It was the sky because there weren't any ceilings anymore. Only ceiling fans. I spun one into an office chair, and sat down.

"You've been working too hard," Audrey said, from where she lay, stretched out on the couch, like a patient, but she was in the white lab coat.

"I'm not a psychologist," she corrected me. "You're making perfect sense all the time. It is only when you stop that you stop making sense."

I was unsure about if this was true. Earlier, it had seemed to me, she could lie the spots off a dog. "But there are no lies." She said it again, for the first time. It was only then that I remembered.

"You haven't thought this way before."

There was a door in front of me, with a large B4 on it. I opened the door, and it was by far the worst thing I had ever seen. It was a straight, continuous line of things and people all making perfect sense up until the present moment, which was me. I slammed the door.

"There, isn't that better now?"

"No, it isn't. Now is never better."

"That is the argument that you must learn to lose."

"But I don't know how to lose arguments."

"You're losing one right now. In fact, it isn't even an argument. It's an aren'tgument."

"We ought to go back to having our first kiss again," I suggested, since we were inside of a fishbowl, and it was always my turn to suggest things when we were inside of fishbowls.

"I'm not kissing you for the first time until you've had a bath," she insisted.

"But we bathed each other already, when we were young, when we were each other's babysitters!"

"Bathe again," I told her. "You stink now, because I have been using you."

"That isn't my fault. Quit being such a man all the time."

When she told me this, I realized what I'd been doing, so I chopped my penis up into little bites, and cooked it in a soup for me to eat. It was delicious. It was also no longer last Tuesday.

"Finally!" she said. "You are a man!"

She burst into exclamation points. I decided to run with that. There were parachutes, when I ran. They burst open all around me. Friends were dangling from them. Had I caused this warzone that they were all trying to escape?

One of them burst into flames beside me. It was Audrey. "No," she said. She divided into two numbers, one, and three, and then she was no longer thirteen, and I was no longer having a cup of tea with her. I supposed that I would be all right with that.

"Are you intending to be surreal, my dear?" she was a clock when she said this. I disagreed.

"You are trying to be surreal. Why else would you be a clock?"

"How should I know? You're the one who fell in love with me."

I hated when she brought that up, because it made the floor erupt underneath me into a volcano of instability, because it made the floor be made out of things that I had said to her before, and I, I feared, could lie the spots off a dog if I had wanted to, but I didn't want to.

"I love you," I insisted.

"Then stop misquoting me."

I burst into music notes, so I wouldn't have a mouth anymore. Mouths weren't important yet; we hadn't yet swum through the ocean preceding our first kiss.

"You're so fixated on that moment," she said. "Just do it already."

"But I did it three years ago."

"So move on."

I thought that she might be suggesting something about my theology, but I could sense the disagreement in her eyes. "You aren't significant enough to be made of theology yet. You still haven't even mastered being made out of romance. You were barely made out of children."

"But I don't like children."

"Because you don't like you. Children are messy and inconvenient because you are."

I was tired of her sounding so much like me, so redundant, so I ate her.

47

"Oh please," she said, rolling her eyes at me. They stopped in front of me on the table. "Pick them up. Roll them like dice. It doesn't matter."

"I always did like dice," I told her, thinking of the furry ones that people hang from car mirrors.

She looked at me. "You don't have a mirror, because you don't have a car, because you don't have a life." She was being very frank. I just couldn't tell what she was being frank about.

"Why don't you try making a little bit of sense?"

She spiraled inward, then, made out of coins. "Are you trying to be a metaphor for greed, my love?"

"Love and metaphors both seem greedy, from a distance."

She slapped her hand on the checkered tablecloth. "Get a hold of yourself, Dream!"

She was becoming another person. I had to kill her, again. It was tedious.

There had been four or five women, maybe six now, I suspected. I had met them all on trains and planes and buses. They'd all been wearing red. I was a murderer.

The airplane I was on – why was I on an airplane that hadn't taken off yet? – turned around and said to me, "I'm still Audrey, so you haven't finished making me you yet."

"Don't be silly, dear. I'm a murderer. I've taken you out to dinner, what is it?"

"Nine times now. I haven't been counting."

Her level of not caring was such that her limbs turned into cats and wandered off of her in opposite directions.

"You couldn't kill me if you tried," the cats dared me.

I wondered why I was wearing a dog catcher's outfit. I wanted to give up, but I didn't. I knew that, deep down, I only APPEARED to be woefully inept. The cats could be caught. I was catching them right now.

"The dog catcher's costume is just a ruse," she told me, when she was the net in my hand. "You aren't succeeding, because you're doing things that aren't success."

"For the love of God!" said the woman across from me on the bus. "Quit staring at me!"

"I'm so sorry, madam. I apologize. I mistook you for someone else."

The teasing smile appeared back on her lips. "Oh, but I am someone else."

"Stop toying with me," I suggested. I suggested it loudly, but it still didn't manage to be more than a suggestion.

She laughed heartily. "Then stop being a toy!"

It was unfortunate, because I then perceived the bleak world that she was offering me, one in which I was not a toy.

I had no sooner rejected the notion of not being a toy any longer, than a large dog came bounding up, clamping me in its jaws.

"You aren't being eaten," Audrey reassured me. "You're being loved. It's almost the same."

"It doesn't feel almost the same," I grumbled.

Grumbling made me turn into garbage. I had fallen a long way, and I had landed in myself, and I was a very large garbage dump.

"You have to uncomplain. It will be a very long time before you can think your way out of this one."

"How do I do it?" I pleaded.

"You can start here, and think a grateful thought for each complaining thought you've ever had, until you've unhad them all."

"But how will I know when I've unhad them?"

"When you're standing here, talking to me, right now."

So we took hands with each other, and walked away. I grew tired of walking for such a long time, so we started to swim. Instantly, my limbs were refreshed. "Now, this is rather like flying," I commented.

"That's because you are flying, silly."

I looked down. When I looked down, I recalled that I didn't know how to fly, and that that was an unfortunate thing to not know how to do, at 34,000 feet up in the sky.

"You really must stop not knowing things," she told me. She turned herself into a school with a very good space program in an attempt to rectify this grievous inadequacy of mine.

I stomped my foot once, and I could speak fluent German. "I think that'll be enough of school," I said, possibly in German.

The school turned into dinosaurs, and they wandered off into outer space, and I wondered off after them. I saw myself coming from a long ways off. He was a sprightly, skinny fellow, with a mop of brown hair, not entirely unlike that fellow in the lab.

"You've spilled something on yourself," Audrey said gently. "It appears to be knowledge. Would you like me to wipe it off?"

"Oh, you can't, can you? It'll stain, won't it? Is it permanent?"

"Knowledge usually is."

She was right. I hated it when she was right, but I also loved it, because I had chosen her, and therefore, if she was right, I was right. Because I'm a man. And men make choices.

"You are not a man unless you are also a boy, last Tuesday."

"But is it always going to be last Tuesday?" I wondered aloud to her, and up at her, because she was cradling me, because, and I was only just now becoming aware of the fact, I was a baby.

"It always was." She said this like I ought to know what she was talking about.

"But you DO know," she said, in that same voice. "And don't tell me not to talk in capital letters."

"No, no, my dear, you are quite right. It is I who should stop speaking in capitol letters." I hadn't noticed until then that I was merely being a pile of paper's on a politician's desk.

"Most people think that they are just a pile of papers on a politician's desk at some point or another, my dear." She was calling me 'my dear' because she was my wife at that point. I had the strangest sensation that I was my own wife, but that couldn't possibly be, because then that would mean that I understood myself.

Suddenly, I was alone. I wanted to, imagined, the act of calling out her name, as either Audrey, or Fidelity. I waited for her to poke her head back around the corner at me. There were no corners, but there would be.

"There are always corners," she said, from behind one, when one appeared. "You're only outside until you realize that there's no such thing as outside. So come it."

I came in, because as it turned out, the corner was a doorway. I realized that I had been cold.

"If you're cold, it's because you're a detective," she decided. "You're a detective coming in out of the cold, on a blustering, wet night."

"I ought to have a British accent, too," I reasoned with her, as she straightened my collar, there by the fireplace.

"Yes," she said. Then she threw me into the fire place. "I decided that I was finished with you being a detective."

"But why do you get to decide?" I said, from the pile of ashes that I now was.

"Because it's already happened." The way she said that, 'it's already happened' made me feel peculiar. There was a floor taking shape around a hole that we were falling down, and I landed on it when I came out of the top of the hole.

"Wasn't that hole a whole lot of fun?" Audrey enthused. I didn't know what she had to be so happy about. A lot of things had clearly just happened, and I didn't know what any of them were.

"Well, there was World War 2," she explained. "You visited there. Then you visited your father. You and your father attended a concert by Elvis Presley, where your father met your mother."

"But my father and mother did not meet at an Elvis concert," I protested. "Elvis is dead."

"So are the things in the hole, which might or might not have happened."

"But I know perfectly well which things have happened and which haven't!" I said, and she walked inside of my mouth. "There you go then," she said. "Just like that. I'm all the medicine you need."

I digested her. If it was true, that she was just herself, and not a string of people whom I had met on the subway and killed, then I was a cannibal, but I was probably a cannibal either way, because I had bitten my toenails as a child.

"Admit it," she said. "You still bite your toenails."

"I don't have to admit anything to you." I knew that I didn't, because she was a tortoise shell, and the tortoise was walking away from me. Unfortunately, it had a planet on its back, and I happened to live on that planet.

"Audrey, wait!" I said.

And she said, "For the last time, or at least for the last time this sentence, my name is Fidelity."

"Fidelity," I said, out of breath. "We've still got a long way to go."

"You seem exhausted," she told me, producing a hotel from her left breast pocket. "Why don't you just disregard everything that's just come out of my pocket, and run a marathon?"

I did not think that this was good advice, but simultaneously noticed that, once again, I was not me. I was just sitting inside of someone else's head. This time, he was fat. He was in front of a tv. He exploded. The tv ate the little exploded bits, and spit them back out into the chair, where I was unfortunately still seated, inside of the man's head. The tv continued to explode, consume, and regurgitate him.

"Fidelity," I said. "Will you kindly lend me a hand?"

She took off her left hand, and handed it to me. I slapped myself in the face.

"I don't like being used as an implement of your own torture," she commented.

"Whyever else would you be here?" Women were here to torture me. I knew this.

"I'm just here because it's your day off, and you are buffooning around on trains, when really, you could be getting somewhere."

"But I am somewhere. Journeys on trains are somewhere."

"You have to unimagined the train. You have to unimagined the things that you are inside of."

She disappeared again when she said that, leaving me alone in a big , black thing. I thought that it was outer space.

"You thought wrong," she said. She was just a voice, by now. "This is inner space. And I can have a body any time you please."

She was standing on a pile of bodies, then, in front of a wrecked subway car. She was crying, but I knew it was an act.

"Did I cause this wreck?" I inquired of her.

"Of course you did," she sobbed. "You imagined all these people, but luckily" – and here she seemed to cheer up very much – "you've imagined them precisely at the moments of their deaths."

All of the dead people got up, then, moving backwards, becoming cleaner, and rocks and bits of wrecked subway car went back into place.

Their lives flashed before their eyes, until they were back behind the bars of their cradles, and then they crawled back up into their mothers' wombs, and there, they got unborn.

"The rules have broken you," she pointed out.

"I don't suppose you know whether or not that's all right?"

"If anything is all right, then everything is all right."

"So, nothing, then?"

"Nothing. Move on."

"Move on, from the fact that nothing is right, at all?"

"But don't you see?" She had recently become a flock of geese, so no, I didn't see. "There are so many lefts!"

She flew off to the left. I was getting a bit worn out with all that, so I slowed my swim to a walk.

"Press on!" she said.

"That's what you said before," I puffed. "Back when we were in my apartment."

"But we're still in your apartment. Stand up."

I stood. We were on a bus. We hadn't met yet. I glanced at the book in her lap, which was face down. I hadn't commented upon it. Nor had I murdered her. Of course, I was relieved.

"Are you sure that relief is the correct choice?" The voice was hers, but it was coming from the inside of the book on her lap.

"Relief is a feeling."

"Choice is also a feeling." She had become a tortoise again, but it was stuck on its back. I decided not to help her up, for once.

"I'm just going to let you flounder there, until you start making some kind of sense."

"All right, all right," she said, her turtle form collapsing into coins. I threw them into a fountain, so that I could make wishes. All of the wishes were for things that I knew she would have wanted. As each coin hit the water, it turned into a butterfly. All of the butterflies were drifting away, and I was concerned about that.

"I feel that I ought to congratulate you on your unprecedented accomplishment in the realm of stupidity," one of the butterflies said. "But I really can't be bothered."

I was stirred by her disregard. I was stirred, like a pot, that she was cooking in. And she was cooking me so well.

I decided to chase this woman, so long as she would allow herself to chased.

"You are looking at me from inside of a cage, still. Release the chimps."

She had, perhaps, told me this before. But that was before I had stomped my foot and learned German, because clearly, she must have said it in German, when she said it before.

I burned the laboratory walls down, revealing a jungle. Hmm, I said, inside of myself.

"Wait. Before you finish that thought, say it out loud, to me."

"That's probably not a good idea, maybe not, you know, your best idea."

"But it's the idea that you're constantly living by."

"I would like to apologize."

"You would like to – but it would be pointless. Moving on."

I could not tell why the release of the chimpanzees, in and of itself, was not satisfactory to her, but I nonetheless followed them in to the jungle. The process was mysterious to me, but I embraced it. The chimpanzees were now gorillas. I was convinced that this change ought to mean something to me, something about the thesis paper I'd been intending to compose?

"You've already done that, dear," my wife reminded me. She was there, by my side, following the gorillas into the jungle. It was night time.

"When did it stop being day time?" I asked her.

"It's still day time. It's just day time somewhere else. Keep following the gorillas through the jungle, and eventually, they're bound to turn up where the sun rises."

She was correct. I tore the sun rise apart when we got there, and shrunk it down, and separated it out, until there was nothing left but a few puddles of color, on a pallet, next to some brushes. I was wearing the smock again. I should never have entertained the idea of not entertaining ideas.

My toes left my feet at that most inauspicious moment, because they had secret things to attend to. I was therefore, temporarily, left without passage, and turned my thoughts inward. There had been something about a book, and either I, or my wife, who was with me in the jungle, whom I had not yet met, were going to be discussing the contents of that book when we did meet.

"We're meeting right now," she said, excitedly, pointing to a clearing up ahead. And there, through the trees, I could almost see us. We were shadows. There were streets. She was running, at first, away from me, down a dark ally, but then, circling back around, she was running toward me, through a field of flowers, and field mice. The field was actually made of field mice, at that point. The grass just stopped being grass, and started being mice, and the mice dissolved into train tracks, and I was lying in an empty subway station, and it was night time. I was holding both of her hands.

"Have we married yet?" I asked her.

"Leave me alone. I know you too well."

I left her. But then I wasn't me anymore, because I only existed as a man in her imagination. I wasn't real. Was I?

"Oh, you're real," she sneered. There was an awful lot of realness, going on, suddenly. There were lights, the kind that police use.

"Yes," she said. "But what else is red, and blue, and noisey?"

"A dying baby!" I exclaimed. We immediately set about saving the baby's life without further discussion. The baby grew up. The baby was me. This grown-up-baby-version of me, thankfully, split itself in twain and one half was vacuumed up into outer space, and the other was washed to a future at the bottom of the blackest sea.

"You are every alternative version of yourself," she commented. She was always 'commeting' and 'noticing' things, in a way that made me suspect that she might be a fictional character.

I decided to exit the train immediately. "But you aren't on a train. You never were in the first place. This is a bus," she said, chasing me through the doors of the train, across the platform, and onto a bus.

"It's a bus because you MADE IT A BUS," I shouted at her.

This was the first time I had ever shouted anything.

"You're still only suggesting things very loudly, for the most part. I'll let you know when you've shouted something good and proper."

I decided to become a vacuum cleaner. I tidied things up into nothingness, assuring myself that, in the end, I would like cleanliness, because my mother had.

"You are going to be in a great deal of trouble," Audrey said – I know her name isn't Audrey, but I'm calling her that, for reasons – "if you don't drink a whole lot, and very fast."

She said that to remind me that I'm made out of water. And things made out of water must always be cannibalizing each other. So I drank the water. Seas of it.

Sea lions came out of my stomach, through my belly button. They said things to me.

"Rudolph the Red Nosed Reindeer is a poor way of seeing yourself. You're not identified by your shortcomings, or the fact that you use those shortcomings to be longcomings. You are just indentified by you. There is no 'red nose.'"

"Except that there is."

"Yes, but the thing about you being you is," the sea lion said, adjusting its bowler derby hat, "you may distract yourself from parts or into parts until you distort yourself into a new you."

"Is this a good thing?" I asked, to the sea lions around me.

They shrugged, and burbled of, back into the ocean, which was leaking out of my eyes. Don't ask us, they seemed to say. We're just sea lions.

I sat down, because I didn't understand.

"You just need to spend a little time as a cartoon character," I said to Fidelity. "You look great as my wife. But you would look even better as a cartoon."

"Let's both be cartoons." I loved the way she'd embrace my ideas, just like they were hers. It almost felt masturbatory.

"Nonsense," she said. "If it were masturbatory, there wouldn't be two of us."

"I'm not sure that there are," I commented. SHE was usually the one who 'commented.' I began to suspect that I was in a book.

"Oh, but you are in a book, darling," she said, through a cartoon mouth. The cartoonishness spread across her face, flattening her out. I tipped my head back, and allowed myself to become two dimensional as well.

"Everything seems terribly high up here, and colorful," she said.

Colors aren't terrible. I knew this, but I didn't say so, because I thought that she probably knew it, too. Colors are nothing to be afraid of. Unless, by being afraid, you mean being respectful.

I was suddenly in a hall of colors. The call was round, and high, and they were seated around chairs. It was dark, in the way that only imposing silences can be dark. The colors were discussing my fate. They knew all of the things that I had known, but they also knew all of the things that I would know eventually. They were the things that I would know eventually.

The table that the colors were sitting around was not a dining table, but a dissecting table. I tied myself down on it. Fidelity stood over me. She held a knife.

"You kept killing me, back there," she said. "Not that I minded. But I thought I might return the favor."

"You can if you want," I said. "But you could always let the colors do it for you."

"The colors don't do things. They are things."

"That's the same," I argued.

"But this isn't an argument. This is an aren'tgument."

I let her dissect me, piece by piece. I did not expect her to find anything of interest.

"Except for this," she said, and I mouthed the words at the same time, because I knew she was going to say it, just like that, when she found it. And she held it up to the light, my heart, pumping stupidly, and she put it in her own chest.

"I was afraid of that," I told her.

I was in the head of the man with the brown eyes again. I was across from her. We were on a train, and a bus, and in a café, and on the floor of the kitchen at home, but these places bored me, so I looked around to see where else we were, ignoring the jungle as well, because I was tired of that. I was tired of outer space, too. And inner space.

"Go get yourself a new name," we told each other, and walked in opposite directions.

It was then that I realized: an idea was having me.